My Counting Trip to the Zoo

by Amy Ayers

Copyright © Gareth Stevens, Inc. All rights reserved.

Developed for Harcourt, Inc., by Gareth Stevens, Inc. This edition published by Harcourt, Inc., by agreement with Gareth Stevens, Inc. No part of this publication may be reproduced or transmitted in any form or by any means, electronic or mechanical, including photocopy, recording, or any information storage and retrieval system, without permission in writing from the copyright holder.

Requests for permission to make copies of any part of the work should be addressed to Permissions Department, Gareth Stevens, Inc., 330 West Olive Street, Suite 100, Milwaukee, Wisconsin 53212. Fax: 414-332-3567.

HARCOURT and the Harcourt Logo are trademarks of Harcourt, Inc., registered in the United States of America and/or other jurisdictions.

Printed in China

ISBN 13: 978-0-15-360220-7
ISBN 10: 0-15-360220-1

7 8 9 10 0940 16 15 14 13
4500409986

Aunt Nina took me to the zoo.

I like the zoo.
It is a fun place.

The zoo has many animals.

We counted all day.
1, 2, 3!

We saw 2 owls.

We saw 7 deer.

There were more deer than owls.

We saw 6 seals.

We saw 3 bears.
We saw fewer bears than seals.

We saw 8 turtles.

We saw 11 bats.
There were more bats than turtles.

Aunt Nina and I got hungry.
We ate lunch.

We ate near a pond.

Then we saw 14 ducks.

Next we saw 9 fish.

We saw fewer fish than ducks.

We saw 4 chimps.

We saw 10 penguins.
We saw more penguins than chimps.

We saw 20 sheep.

We saw 18 storks.

We saw fewer storks than sheep.

Then we looked for snakes.
We saw 0 snakes.

We saw 4 wolves last.
We saw more wolves than snakes.

I took lots of pictures.

I will make a scrapbook.
I will show Aunt Nina!

Glossary

fewer There are fewer bears than wolves.

more There are more ducks than bats.

Photo credits: cover, p. 20 © Gary D. Landsman/Corbis; p. 2 © Nik Wheeler/Corbis; p. 3 © William Manning/Corbis; p. 4 © Lon C. Diehl/Photo Edit; p. 5 Adam Jones/Visuals Unlimited/Getty Images; p. 6 U.S. Fish and Wildlife Service; p. 7 Raymond Gehman/National Geographic/Getty Images; p. 8 © Wolfgang Kaehler/Corbis; pp. 9, 24 (top left) © Mark Newman/FLPA; p. 10 © Jim Merli/Visuals Unlimited; pp. 11, 24 (bottom left) © David Hosking/FLPA; pp. 12, 23 Russell Pickering; p. 13 © Inga Spence/Visuals Unlimited; pp. 14, 24 (bottom right) © Klaus Hackenberg/zefa/Corbis; p. 15 © Warren Morgan/Corbis; p. 16 © Mary Kate Denny/Photo Edit; p. 17 © Galen Rowell/Corbis; p. 18 Nina Buesing/Stone+/Getty Images; p. 19 © Eric Wanders/Foto Natura/FLPA; p. 20 © Gary D. Landsman/Corbis; pp. 21, 24 (top right) Guy Edwardes/The Image Bank/Getty Images; p. 22 Jaimie D. Travis/DK Stock/Getty Images.